I0794862

Review of the U.S. Army Corps of Engineers Restructured Upper Mississippi-Illinois River Waterway

FEASIBILITY STUDY

Committee to Review the Corps of Engineers
Restructured Upper Mississippi River-Illinois Waterway
Draft Feasibility Study

Water Science and Technology Board
Division on Earth and Life Sciences

Transportation Research Board

NATIONAL RESEARCH COUNCIL
OF THE NATIONAL ACADEMIES

THE NATIONAL ACADEMIES PRESS
Washington, D.C.
www.nap.edu

THE NATIONAL ACADEMIES PRESS **500 Fifth Street, N.W.** **Washington, DC 20001**

NOTICE: The project that is the subject of this report was approved by the Governing Board of the National Research Council, whose members are drawn from the councils of the National Academy of Sciences, the National Academy of Engineering, and the Institute of Medicine. The members of the committee responsible for the report were chosen for their special competences and with regard for appropriate balance.

This study was supported by Contract No. DACW17-03-C-0003 between the National Academy of Sciences and the Department of the Army. Any opinions, findings, conclusions, or recommendations expressed in this publication are those of the author(s) and do not necessarily reflect the views of the organizations or agencies that provided support for the project.

International Standard Book Number 0-309-09133-0 (Book)
International Standard Book Number 0-309-52965-4 (PDF)

The Upper Mississippi River-Illinois Waterway Draft Feasibility Study: Interim Report is available from National Academies Press, 500 Fifth Street, N.W., Lockbox 285, Washington, DC 20055; (800) 624-6242 or (202) 334-3313 (in the Washington metropolitan area); Internet, http://www.nap.edu

Cover design by Michael Dudzik, the National Academies Press. Photos courtesy of John Boland, Johns Hopkins University and U.S. Army Corps of Engineers websites: http://www.mvr.ausace.army.mil/navdata/Pics/12(92).jpg and http://www.mvr.usace.army.mil/mvrimi/omni/external/rpt04/RPT04Home.asp. Photo on back cover courtesy of Jon Duyvejonck, U.S. Fish and Wildlife Service.

Copyright 2004 by the National Academy of Sciences. All rights reserved.

Printed in the United States of America

THE NATIONAL ACADEMIES

Advisers to the Nation on Science, Engineering, and Medicine

The **National Academy of Sciences** is a private, nonprofit, self-perpetuating society of distinguished scholars engaged in scientific and engineering research, dedicated to the furtherance of science and technology and to their use for the general welfare. Upon the authority of the charter granted to it by the Congress in 1863, the Academy has a mandate that requires it to advise the federal government on scientific and technical matters. Dr. Bruce M. Alberts is president of the National Academy of Sciences.

The **National Academy of Engineering** was established in 1964, under the charter of the National Academy of Sciences, as a parallel organization of outstanding engineers. It is autonomous in its administration and in the selection of its members, sharing with the National Academy of Sciences the responsibility for advising the federal government. The National Academy of Engineering also sponsors engineering programs aimed at meeting national needs, encourages education and research, and recognizes the superior achievements of engineers. Dr. Wm. A. Wulf is president of the National Academy of Engineering.

The **Institute of Medicine** was established in 1970 by the National Academy of Sciences to secure the services of eminent members of appropriate professions in the examination of policy matters pertaining to the health of the public. The Institute acts under the responsibility given to the National Academy of Sciences by its congressional charter to be an adviser to the federal government and, upon its own initiative, to identify issues of medical care, research, and education. Dr. Harvey V. Fineberg is president of the Institute of Medicine.

The **National Research Council** was organized by the National Academy of Sciences in 1916 to associate the broad community of science and technology with the Academy's purposes of furthering knowledge and advising the federal government. Functioning in accordance with general policies determined by the Academy, the Council has become the principal operating agency of both the National Academy of Sciences and the National Academy of Engineering in providing services to the government, the public, and the scientific and engineering communities. The Council is administered jointly by both Academies and the Institute of Medicine. Dr. Bruce M. Alberts and Dr. Wm. A. Wulf are chair and vice chair, respectively, of the National Research Council

www.national-academies.org

COMMITTEE TO REVIEW THE CORPS OF ENGINEERS RESTRUCTURED UPPER MISSISSIPPI RIVER-ILLINOIS WATERWAY DRAFT EASIBILITY STUDY[1]

JOHN J. BOLAND, *Chair*, John Hopkins University, Baltimore, Maryland
PATRICK BREZONIK, University of Minnesota, Minneapolis
ROBERT K. DAVIS, University of Colorado, Boulder (retired)
LEO M. EISEL, Brown and Caldwell, Golden, Colorado
STEPHEN W. FULLER, Texas A&M University, College Station
GERALD E. GALLOWAY, Titan Corporation, Fairfax, Virginia
LESTER B. LAVE, Carnegie Mellon University, Pittsburgh, Pennsylvania
KARIN LIMBURG, Syracuse University, New York
ELIZABETH A. RIEKE, U.S. Bureau of Reclamation, Carson City, Nevada
RICHARD E. SPARKS, University of Illinois, Urbana-Champaign
SOROOSH SOROOSHIAN, University of California, Irvine

NRC Staff

JEFFREY W. JACOBS, Study Director, Water Science and Technology Board
JOSEPH MORRIS, Senior Staff Officer, Transportation Research Board
ANITA A. HALL, Senior Project Assistant, Water Science and Technology Board

Editor

FLORENCE POILLON

---

[1] The activities of the Committee to Review the Corps of Engineers Restructured Upper Mississippi River-Illinois Waterway Draft Feasibility Study are overseen and supported by the National Research Council's Water Science and Technology Board (lead) and the Transportation Research Board.

Preface

The U.S. Army Corps of Engineers has a long history of managing navigation, floods, and other water-related issues on the Upper Mississippi and Illinois Rivers. A recent chapter in that history is the problem of waterway congestion at several locks on the lower portion of the Upper Mississippi River. Barges and other vessels must often wait in queues to transit these locks, with delays ranging from several hours to several days. The Corps has studied this problem and its possible solutions since the late 1980s, producing a draft feasibility study in 2000 and an interim report on a restructured feasibility study in 2002. The restructured feasibility study is scheduled for completion late in 2004.

This committee was convened to review and provide advice on the most recent phase of the Corps' analytical efforts. The extended duration of the Corps' study is a result of the issue's considerable complexity and of Corps efforts in correctly framing the issue as a systems-level problem, as well as the sometimes conflicting economic and cultural values related to managing this huge and valuable river system.

This report follows the committee's initial meeting, in which discussions with the Corps, several stakeholders, and among ourselves led to several first impressions and recommendations regarding the Corps' ongoing restructured feasibility study. Although our committee has much to learn about the Upper Mississippi and Illinois Rivers, waterway traffic, and river ecology, we have been impressed by the complexity of these issues and their importance for local people and communities, the nation at large, and the role of U.S. grain production in the world. We look forward to learning more as our investigation proceeds.

We thank the Corps' professional staff for providing an abundance of relevant information in a timely and helpful manner. At our September 2003 meeting in Washington, D.C., we spoke with Brigadier General Donald Riley, Colonel Duane Gapinski, and Corps study team members Denny Lundburg, Kenneth Barr, and Richard Manguno, among others. We have also received excellent advice and assistance from William Dawson and Richard Worthington at Corps Headquarters in Washington. In speaking with these Corps staff and in learning more about the study, we have come to respect the pressures under which they are operating and appreciate the open exchanges of ideas and information we have had with them. We also appreciated the opportunity to speak with and learn from Christopher Brescia, Scott Faber, Rick Moore, Holly Stoerker, and Sander Toth at our September 2003 meeting.

This report was reviewed in draft form by individuals chosen for their diverse perspectives and technical expertise in accordance with the procedures approved by the National Research Council's (NRC's) Report Review Committee. The purpose of this independent review is to provide candid and critical comments that will assist the institution in making its published report as sound as possible and to ensure that the report meets institutional standards for objectivity, evidence, and responsiveness to the study charge. The review comments and draft manuscript remain confidential to protect the integrity of the deliberative process. We thank the following for their reviews of this report: Phillip Baumel, Iowa State University; Stanley A. Changnon, University of Illinois; José A. Gómez-Ibánez, Harvard University; James Heaney, University of Colorado; Leroy Poff, Colorado State University; Jerald L. Schnoor, University of Iowa; and Leonard Shabman, Resources for the Future.

Although the reviewers listed above have provided many constructive comments and suggestions, they were not asked to endorse the conclusions or recommendations, nor did they see the final draft of the report before its release. The review of this report was overseen by Walter R. Lynn, Cornell University, appointed by the Division on Earth and Life Studies, and by Frank H. Stillinger, Princeton University, appointed by the NRC's Report Review Committee, who were responsible for making certain that an independent examination of this report was carried out in accordance with institutional procedures and that all review comments were carefully considered. Responsibility for the final content of this report rests entirely with the authoring committee and the institution.

The committee is particularly grateful for the hard work and wise counsel of the prior NRC committee that reviewed the Corps' 2000 draft feasibility study and was chaired by Lester Lave. We have also benefited from the constant assistance and support of NRC staff members, including Stephen Parker, Steve Godwin, Joseph Morris, and Anita Hall. Finally, we are most indebted to Study Director Jeffrey Jacobs for the smooth and effective conduct of this review, as well as the considerable achievement of making 11 authors seem to speak in a single voice.

John J. Boland, *Chair*

Contents

Executive Summary

This is the first of three reports from the National Research Council (NRC) Committee to Review the U.S. Army Corps of Engineers Restructured Upper Mississippi River-Illinois Waterway Feasibility Study. The committee was assembled in response to a request from the Corps of Engineers (Corps) for the National Research Council to review the feasibility study. The committee's assignment is to provide advice on the Corps' Restructured Upper Mississippi River-Illinois Waterway (UMR-IWW) System Navigation Feasibility Study. To this end, the committee is reviewing several Corps documents that explain the analysis within the feasibility study, including the Corps' July 2002 Interim Report. Since a previous NRC (2001) review, the nature of the Corps feasibility study has broadened beyond the need for transportation improvements, and the restructured feasibility study takes a more holistic approach toward considering the relations between environment, navigation, and the floodplain. Although the committee has the discretion to determine appropriate topics for its review, the review was expected to include several topics on which the prior committee had commented and for which the Corps has proposed responses in its restructured study plan (Chapter 1 contains the committee's full statement of task).

This report was written after the committee's review of several documents provided by the Corps and after a briefing by the Corps in early September 2003. This report represents the committee's initial impressions of the restructured feasibility study. It was written with the intent that the

Corps could consider its findings and recommendations as the agency refines its feasibility study. The report comments on the following components of the Corps' feasibility study: (1) a spatial equilibrium model and the ESSENCE model, both developed to help forecast future grain shipment levels, transport modes, and costs on the UMR-IWW; (2) managing waterway congestion on the UMR-IWW; (3) integrated systems planning; and (4) decisions, implementation, and institutions. A second, more comprehensive report will be issued in 2004, in which the committee will comment on a Corps draft feasibility report (currently scheduled for publication in April 2004). Finally, a third report will review the Corps' responses to advice from the NRC Upper Mississippi River-Illinois Waterway studies (i.e., the reports of this committee and the NRC 2001 "Phase I" committee).

SPATIAL PRICE MODEL AND ESSENCE

The Corps has worked toward developing a spatial price model for the feasibility study, which could create forecasts of spatial variations in grain shipping costs, modes of transport, trends in grain processing, and waterway traffic levels on the Upper Mississippi River-Illinois Waterway. The Corps has applied and been developing various models in this effort, including a traditional Corps Tow Cost Model (TCM) and an ESSENCE model.

The TCM was previously rejected by the Corps, a decision with which the NRC 2001 committee concurred. That committee also recommended that results from the ESSENCE model should not be used in the feasibility study. However, the Corps has reversed its decision on the use of the TCM and continues to use the ESSENCE model. This committee has not been presented with detailed information on the TCM, and it looks forward to learning more about the TCM and the decision to use it in the feasibility study. Fundamental conceptual flaws in the ESSENCE model render it unfit for use in the feasibility study. The Corps should either accelerate the development of a full spatial price model or adopt an alternative modeling approach that avoids the limitations of the ESSENCE model.

Demand Forecasts

Forecasts of future grain exports are being used in the feasibility study, and credible forecasts are essential to sound investment decisions on the UMR-IWW. Of five future scenarios developed for the feasibility study, four reflect increasing grain exports. These forecasts, however, are inconsistent with the past 20 years of relatively steady export levels. The committee reserves judgment on the plausibility of these forecasts and looks forward to discussing the rationale for the projections with the Corps' consultants at its next meeting.

MANAGING WATERWAY CONGESTION

The current system for managing UMR-IWW traffic operates largely on a first-come, first-serve basis. It is possible that new traffic management measures would reduce congestion in a cost-effective manner. Unless an efficient system for managing waterway traffic is in place, it is not possible to evaluate the benefits of lock extensions. Nonstructural measures such as scheduling systems, systems of tradable arrival slots, or a contingent fee—as challenging as their implementation may be—could be implemented instead of extending locks or could be used in combination with lock extensions. Furthermore, such traffic management measures would be necessary to address reductions in capacity that would occur in the event that lock extensions were constructed. The Corps should proceed as soon as practicable toward developing and implementing any feasible nonstructural measures to help alleviate waterway traffic congestion.

INTEGRATED SYSTEMS PLANNING

The Upper Mississippi and Illinois Rivers and their floodplains support a variety of activities other than navigation, including boating, recreational and commercial fishing, hunting, camping, and sightseeing. These activities are important to both the economies and the quality of life in dozens of communities along the river, including several cities. The Corps should thus, to the extent feasible, consider these and other factors such as water quality, flood damage reduction, and sediment transport in order to reflect a more holistic approach to dealing with the diverse management issues in the UMR-IWW. The Corps should also implement adaptive management concepts and approaches throughout the planning process. Finally, the Corps should acknowledge the potentially high cost of the proposed restoration efforts and the necessity of conducting ecological investigations in tandem with navigation studies.

The Corps is seeking to integrate the relationships and functions of floodplains. This integration is manifested primarily in the form of ecosystem restoration projects in floodplain areas along the rivers. Information presented to the committee to date about prospective restoration projects has contained few details, making it difficult to say how these projects might affect the floodplain. It is also not clear how far the Corps will go at integrating floodplain concerns into the feasibility study. The holistic approach envisioned in the Corps' study, which integrates the floodplain into operations of the navigation system, is not yet clear. Further comment on issues in the floodplain is thus reserved until additional information is provided to the committee.

DECISIONS, IMPLEMENTATION, AND INSTITUTIONS

The Corps is on an aggressive timetable for completion of the feasibility study. Although there is a need to move forward with the study, that need should be balanced with a similar need for credible and thorough analytical procedures. The Corps should extend its schedule for completing the feasibility study and issuing a "Chief's Report."

There are literally hundreds of prospective projects within the UMR-IWW study aimed toward ecological enhancements. Not all of these projects can be implemented simultaneously, because of both financial and practical limitations. Priority should be given to restoration projects that promise to restore natural processes and those that aim to achieve multiple objectives.

Ecological restoration projects in downstream states, which have smaller federal land holdings than states in the upper basin, require greater cost-sharing responsibilities from local sponsors. This is an analytical concern because the prospective benefits of these cost-shared restoration projects are an important component of the feasibility study. If the projects do not receive funding, this will alter the study's benefit calculations. The Corps should thus identify specific instances in which federal cost-sharing rules are likely to restrict or preclude implementation of environmental restoration projects and nonstructural measures.

1

The Corps of Engineers and the Upper Mississippi River-Illinois Waterway

INTRODUCTION

The U.S. Army Corps of Engineers has been involved in navigation enhancement activities on the Upper Mississippi and Illinois Rivers since the early nineteenth century, when it was charged to remove snags and overhanging trees that impeded navigation. In 1878, Congress authorized the Corps to construct a 4.5-foot navigation channel project on the Upper Mississippi River. In 1907, congressional authority was granted to the Corps to construct a 6-foot navigation channel, and in the 1930 Rivers and Harbors Act the Corps was requested to construct a 9-foot navigation channel project. The 1930 authorization "represented a turning point" (NRC, 2001) in Upper Mississippi River history, because it resulted in the construction of 26 locks and dams that impounded many stretches of the river, creating a series of navigation pools and a 9-foot navigation channel. Completed in 1940, the project supported increases in commercial towboat traffic and initiated changes in ecological structure and processes that continue to affect river ecology and system dynamics.

Upon completion of the lock and dam project and the 9-foot channel, commercial traffic on the river increased steadily. In the year 2000, the Upper Mississippi River carried 122 tons of commercial cargo (USACE, 2002).

With increasing waterway traffic, congestion became a problem at some locks on the lower portion of the Upper Mississippi (just north of St. Louis). One factor contributing to the congestion is that the length of tows on the river has increased over time. Most of the locks in the Upper Mississippi River-Illinois Waterway (UMR-IWW) are 600 feet (their original length), while most tows today push multiple barges that can be nearly 1000 feet. Tows must thus de-couple their barges and transit through the locks in "double lockages," which delays transit times. Congestion at these locks induced the Corps of Engineers to initiate studies to evaluate the alternatives for reducing waterway congestion.

The Corps began its investigations with separate studies of the Upper Mississippi River and the Illinois Waterway in the late 1980s, combining them into a single feasibility study in 1993. The study's analytical complexities slowed its progress. These complexities also led the Department of Defense (DOD) to enlist the National Academies and its research arm, the National Research Council (NRC), to provide an independent review of drafts of the feasibility study.

NRC Committees to Review the UMR-IWW Feasibility Study

In February 2000, the Department of Defense requested that the National Research Council convene a committee to review the Corps' feasibility study. That committee ("Phase I" committee) completed its study in early 2001 (NRC, 2001). That committee was requested to focus its review on the Corps' economic analysis of proposed navigation system improvements, but also to comment on other relevant water resources planning issues. The Corps subsequently "restructured" its feasibility study, with an important milestone being marked by the issuance of a July 2002 interim report (USACE, 2002). According to the Corps' current schedule for the feasibility study, the agency plans to issue its final report and recommendation in a "Chief's Report" in November 2004.

In March 2003, the Corps requested the National Research Council to convene another committee ("Phase II" committee) to review the Corps' progress with its restructured feasibility study. This review was to focus on the Corps' 2002 interim report, as well as various supporting documents produced by the Corps and some of its consultants. The statement of task for the Phase II committee follows:

> The committee will review the Corps' Restructured Upper Mississippi River-Illinois Waterway System Navigation Feasibility Study. The committee will review several Corps documents that explain the analysis within the feasibility study, including the Corps' July 2002 Interim Report for the study. A key document for the committee's review will be a summary of the feasibility study that the Corps will provide to the committee before its first meeting. Since the 2001 NRC report, the nature of the Corps' feasibility study has broadened beyond the need for transportation improvements; the restructured feasibility study has taken a more holistic approach toward considering the relations between environment, navigation, and the floodplain. Given the emphasis on comprehensive river system planning in the restructured study, the committee will provide a comprehensive review of all aspects of the feasibility study, including economic evaluation, environmental analysis, design and engineering, and plan formulation focusing on key study assumptions.
>
> The committee will focus its review on the key study issues, assumptions, and areas of controversy. Although the committee will have discretion to determine appropriate topics for its review, it is expected that the review will include several topics that the prior committee commented upon and for which the Corps has proposed responses in the restructured study plan. These topics include the Corps' decision to replace the ESSENCE model with its Tow Cost Model; the appropriateness of the scenario-based forecasts of barge demand by commodity and how these varied scenarios will be incorporated into the subsequent analyses; how the restructured plan should incorporate the nonstructural alternatives (pricing, scheduling, etc.) into the feasibility analysis; the potential effectiveness of the proposed environmental restoration, its costs, and how

> the cost should be apportioned among the involved parties (federal, state, local, and private); and broad matters related to water resources systems planning and decision analysis.

The Corps requested that this NRC Phase II committee produce three documents: (1) a report of the committee's initial impressions of the restructured interim report, (2) a more comprehensive and detailed report, and (3) a final summary report. The Corps provided several documents to this committee and met with the committee in Washington, D.C., on September 8, 2003, to present and discuss the study. This document is the Phase II committee's initial report, and it provides the committee's initial impressions of the Corps' restructured feasibility study. It is based on review of and deliberations regarding documents and other information provided by the Corps during or prior to the September 2003 meeting. The Corps provided other background documents after the meeting, which the committee has not discussed as a group or discussed with the report authors. Although some of these documents are referred to in this report, the committee reserves judgment about them. Chapter 1 of this report is descriptive and provides background material related to information needs in inland water system planning. Chapter 2 presents the committee's findings and recommendations.

The relationship between the Phase I and Phase II committees merits a brief explanation. The Phase II committee is a distinct activity from the Phase I committee. Its statement of task is different and it is reviewing different (in some cases revised) Corps documents. Membership of these committees is different: the Phase II committee includes only one member who served on the Phase I committee. It should be emphasized that this Phase II committee is not obliged to concur with any of the Phase I committee report's recommendations. In many cases, the Phase II committee uses recommendations from the Phase I committee as starting points for discussion, and there are instances in which this Phase II committee agrees with findings and recommendations from the NRC (2001) report (which are noted explicitly herein). However, this report comes from a distinctly different committee and does not imply full endorsement of all recommendations in the Phase I committee's report. Having said this, the Phase I committee's report (NRC, 2001) is prominent in the Corps' refinements to its feasibility study, and this Phase II committee has thus carefully studied the 2001 document. The following section highlights a few key issues, findings, and recommendations from the Phase I committee report.

Phase I Committee Report

National economic issues such as waterway traffic costs, levels of waterway traffic, forecasts of future grain exports, assumptions regarding ports of grain export, and the ability of shippers to use alternative modes of transportation (e.g., railways) as waterway shipping costs vary, figure prominently in the Corps' feasibility study. To help understand these issues, the Corps developed a spatial equilibrium model for use in the feasibility study. The Corps also developed an "ESSENCE" model used to calculate equilibrium values for barge traffic and economic benefits associated with relieving waterway congestion. The Phase I committee report noted that these models represented conceptual advances over previous Corps efforts in this realm but found that they were characterized by "flawed assumptions and data." Its report concluded that the Corps' "current (September 2000) results . . . should not be used in the feasibility study" (NRC, 2001, p. 3).

Another key conclusion of the Phase I committee was that the UMR-IWW system of locks is not being used efficiently. There is no formal waterway traffic management or scheduling system, and the lack thereof contributes to occasional and random delays. The report noted, "If barge traffic was distributed more evenly, congestion would decrease and shipping costs would fall" (NRC, 2001, p. 3).

The Phase I committee also concluded that the feasibility study had framed the issues of waterway traffic narrowly and that critical topics such as environmental impacts and restoration had received inadequate consideration. This led to a recommendation for "a more comprehensive and integrated assessment of the navigation system's effects on the environment in the UMR-IWW" (NRC, 2001, p. 4). Other important outcomes of the Phase I committee's study included a recommendation to include independent review within the feasibility study, a statement concerning the value of conducting the study and managing the system with an adaptive management approach, and a finding that a contingency figure for the costs of extending locks likely had been underestimated.

INFORMATION NEEDS FOR THE UMR-IWW FEASIBILITY STUDY

This study is using some of the findings and recommendations from the NRC (2001) report as criteria for evaluating the analysis within the Corps' restructured feasibility study. This approach is appropriate given that the restructuring has been conducted primarily in response to the Phase I committee report. In documentation provided to this Phase II committee, the Corps noted instances in which it has chosen not to implement recommendations from the Phase I report (USACE, 2003a). This committee is not evaluating the feasibility study based solely on the extent to which it has adopted the Phase I committee's recommendations. Rather, the feasibility study will be judged by the criterion of whether it is likely to be useful as a basis for well-informed and credible decisions. Many issues addressed in the feasibility study will contain uncertainties despite the best analytical efforts to reduce them. Given the existence of irreducible economic, ecological, and other uncertainties, investment and management decisions cannot be delayed indefinitely in a never-ending search for more information. Comprehensive, sound analysis, however, can highlight key uncertainties, reduce uncertainties in some cases, and provide probabilities of different outcomes, all of which are valuable to planners and decision makers.

Navigation Economics

Forecasting Future Waterway Traffic on the UMR-IWW

Good decisions regarding investments in large civil works projects such as lock extensions on the Upper Mississippi River require some consideration of the future demands for those projects. Corps of Engineers locks and dams are intended to have a useful life of many decades. Planners must thus forecast future demands for these projects' services over decades-long time spans. Such long-term forecasts inherently contain large uncertainties. Creating long-term forecasts that are plausible enough to support long-term decisions is an analytical challenge. Also, although discount rates may diminish the importance of forecasts as one goes farther into the future, credible long-term forecasts are nonetheless essential to sound planning. Flawed forecasts may result in useful projects' not being constructed in a timely fashion or in the construction of projects for which subsequent demand "falls short of" forecast levels and thus fails to justify the investment.

A spatial price model is one means for gauging how future levels of traffic on the UMR-IWW might change in response to a variety of factors such as changes in shipping rates; the availability of alternative transport modes and ports of export; trends in grain processing; and shifts in regional, national, and global patterns of grain supply and demand. The model also would gauge the degree to which grain shipments increase, decrease, or move to alternative transportation modes with changes in the cost of waterway transport, a phenomenon described as the "elasticity" of waterway traffic demand.

Spatial models represent an appropriate method for examining lock congestion on inland waterways. Increases in traffic and lock congestion on the UMR-IWW would be expected to increase waterway shipping rates that link supply regions (e.g., central Iowa) and Gulf of Mexico ports. These increases could be reflected in a spatial grain model, which could make upward adjustments in waterway shipping rates that link various Upper Mississippi River elevator sites to Lower Mississippi River ports. Increases in waterway shipping rates would provide an incentive for shippers to examine alternative markets (e.g., regional corn or soybean processors; poultry and livestock producers) and alternative transportation routes (e.g., rail to the Gulf of Mexico or to the Pacific Northwest). Subsequent analyses could estimate the likely reduction in barge shipping rates that would result from improved lock infrastructure (e.g., mooring, lock extension, lock rehabilitation). The reduced waterway shipping rates could be incorporated into the spatial models and the models solved to determine increases in river grain flow and increases in economic welfare resulting from these investments. Eventually, the Corps would likely wish to develop spatial price models that featured several crops, land resource constraints, and a longer time perspective. Regardless, these types of models represent an analytical building block that could be augmented within the feasibility study (Appendix A provides more details regarding examples of spatial equilibrium models and their applications to the U.S. grain sector).

The Tow Cost Model and ESSENCE

The Corps of Engineers traditionally has estimated waterway traffic levels using a Tow Cost Model (TCM). Documents and briefings provided to the NRC Phase I committee indicated that the Corps had developed a "spatial equilibrium" (ESSENCE) model to explain changes in grain shipping modes and export levels with different levels of waterway traffic costs and congestion. The Phase I committee commended the Corps for this development, noting that it represented important conceptual progress beyond earlier models, such as the TCM: "This system model represents a major advance over previous economics models used by the Corps to forecast barge traffic" (NRC, 2001). The Phase I committee also noted, however, that implementation of the ESSENCE model was inadequate: "The ESSENCE model does not, however, adequately use the more important concepts of the spatial equilibrium model that were advocated in the draft feasibility study" (NRC, 2001).

The Corps has continued its efforts to develop a realistic and reliable spatial price model. The model is not yet completed however, and Corps staff implied that it may be years until it is. Until the model is developed, the Corps has elected to apply both the Tow Cost Model and the ESSENCE model, the latter of which contains two values of a coefficient that characterizes the shape of the demand curve for waterway traffic. One of these values is described as a lower bound and one as an upper bound value. Both of these parameter values are based on the judgment of Corps staff; no data are provided to estimate this critical parameter or justify the values chosen.

After the Phase I report was issued, a federal interagency "Principals Group" was created to guide the development of the Corps' feasibility study. Members of the Principals Group include representatives from the Corps of Engineers, the Department of Agriculture, the Department of Transportation, the Environmental Protection Agency, and the Fish and Wildlife Service. This Principals Group was cited as recommending the use of the TCM to give a common metric to allow the UMR-IWW study to be compared to previous Corps studies: "The Federal Principals Group endorsed the use of existing and accepted economic models while research and development on improved models moves forward . . ." (USACE,

2003a, p. 8). Although this Phase II committee did not review the Tow Cost Model in detail, the decision to use the Tow Cost Model in the feasibility study prompts two observations. One is that it is the committee's understanding that the federal Principals Group was not created to provide advice on technical issues, but rather to provide broad policy guidance. A Corps of Engineers web site, for example, describes the Principals Group as follows: "The National Federal Senior Principals Task Force was established by the Corps of Engineers to provide national-level balance and guidance on important economic and environmental issues to assist in bringing this study to completion" (*http://www2.mvr.usace.army.mil/umr-iwwsns/-index.cfm?fuseaction=home.faq*; accessed November 6, 2003). It is not clear that the Principals Group clearly understood the technical details involved in model selection.

A second observation regards the claim that the Tow Cost Model represents an "accepted" economic model. Although this committee has not yet been provided details of the TCM, the Corps apparently concluded years ago that the TCM was inappropriate for the feasibility study and spent considerable time and resources on developing an alternative. The committee has not been presented with information to suggest that the TCM has been adjusted so that it can provide reliable estimates of the benefits of lock extensions on the UMR-IWW. More information regarding the TCM and the decision to use it in the feasibility study will be presented and reviewed later in the course of this study.

The Phase I committee noted several limitations of the ESSENCE model. The model represents a significant simplification of the spatial equilibrium concept in that it incorrectly assumes that the cost of transport is always proportional to distance. The Phase I report (NRC, 2001) recommended that that the ESSENCE model be recast to remove this and other limitations in order to make it more amenable to use with a spatial equilibrium model, recommending that the Corps should "revise the ESSENCE model, eliminating assumptions that shipment costs are proportional to distance . . ." (NRC, 2001, p. 3). The Phase I committee also recommended that the revised ESSENCE model be implemented by having parameters estimated using empirical data from the UMR-IWW, stating that the Corps should "estimate demand and supply sensitivities to price from studies of current data . . ." (NRC, 2001, p. 3). Price responsiveness is so important to estimating the benefits of waterway improvements that informed judgments about the merit of such improvements cannot be made without careful study of these demand and supply elasticities. The values of these elasticities should be based on empirical demand and supply data for the Upper Mississippi River-Illinois Waterway. It is unlikely, however, that these empirical data can be gathered and adequately evaluated within the schedule the Corps has set for the feasibility study. Furthermore, the present specification of the ESSENCE model does not allow for an independent choice of demand elasticity, whether or not it is empirically determined.

Demand Forecasting

It is not possible to predict grain movements on the UMR-IWW 50 years into the future with anything approaching a high degree of confidence. Nonetheless, some estimates of future conditions are necessary for a project of the scale being evaluated by the Corps. One way to address the problems inherent in this forecasting is to create multiple scenarios of a range of plausible future conditions. The Phase I committee recommended scenario analysis as an option for the Corps, and the present committee commends the Corps for developing this approach for the feasibility study.

If plausible projections for future waterway traffic levels are to be produced, the fundamental structural factors, or "drivers," that affect regional and global patterns of production, consumption, and trade must be identified and explained. A valid approach for scenario analysis would include an explanation of key drivers (e.g., changes in technology, consumer preferences, global climate, trade policy, population growth), as well as significant candidate drivers that may have been omitted from the analysis. A valid approach also would include a reasonable range of plausible futures, supported by explanations for projected changes in key drivers. A set of scenarios that all reflect essentially the same future trend should be cause for suspicion. This would not necessarily imply that all the scenarios are invalid, but it should induce the analysts to seek clear explanations behind the assumptions of the driving forces upon which the scenarios are based. The credibility of the explanations for the drivers would depend largely on their consistency with expert opinions about trends and futures in national and global grain markets. Thorough review by independent experts would help ensure the validity of key drivers and the credibility of the resulting scenarios. Such scenarios will always contain a degree of uncertainty, and uncertainly alone should not justify the delay of investment decisions. However, the magnitude and the potential effects of investments being considered in the feasibility study require scenarios that are consistent with key drivers in global and national grain markets, that are supported by credible model results, and that are consistent with the knowledge of credible and independent experts.

The credibility of models used to create waterway traffic demand forecasts for the UMR-IWW feasibility study should be validated by their ability to reproduce actual historical patterns of waterway traffic. If the application of past values for the selected drivers could reasonably reconstruct patterns of waterway traffic from 1980 to 2003, for example, the modeling approach and the scenarios developed from it would gain credibility. On the other hand, if the approach cannot reasonably recreate past historical patterns, it should be reexamined for key assumptions, data, and algorithms. Testing for ability to successfully "backcast" is the conventional method for validating models for use in generating forecasts or scenarios.

Managing Congestion

The Corps is to be commended for considering the Phase I committee's recommendation to examine the use of one or more tolls to help manage waterway traffic congestion. Even if the proposed lock replacement and extension construction were started today, a decade would pass before its completion. In contrast, there are other forms of nonstructural waterway traffic management options. A congestion toll, for example, requires no construction and could provide immediate benefits. Indeed, during the decade-long construction period, the effective capacity of the waterways will be reduced, resulting in even greater benefits from a congestion toll.

The committee inference from the Corps' 2002 Interim Report is that the Corps has evaluated a congestion toll in terms of a lockage fee that would be *constant* over time. The congestion toll should be a *contingent* fee, however, and should vary with varying levels of congestion: "One market-based system is a congestion toll, where each tow pays for the cost of the delays imposed on other tows" (NRC, 2001, p. 69). Thus, if a tow were willing to wait until there was a vacant lock for its passage, it would have a zero congestion toll. This responsive congestion toll would generate higher benefits than a constant toll of similar magnitude. An aspect of congestion tolls not addressed by the Corps' report is the financial impact of congestion tolls on towboat operators. It is possible to employ deposit-refund schemes to return most toll revenues to the industry, preserving the incremental incentives provided by the toll while reducing the average financial impact on operators and shippers. Other nonstructural traffic management alternatives that could also be evaluated include tradable arrival slots, industry self-help, and the use of switchboats.

Principles and Guidelines: Nonstructural Alternatives

The congestion toll described above is one of several nonstructural alternatives that could be used in conjunction with or as a substitute for structural improvements. Before undertaking a billion dollar construction project that will affect river traffic for a decade and have large-scale effects on the environment and river ecology, it should be determined that the level of construction is appropriate to the level of demand. Traffic on the UMR-IWW is served largely on a first-come, first-serve basis, and there is no scheduling or other management system that attempts to smooth waterway traffic flows. The system is thus not being used efficiently, which led the Phase I committee to note that ". . . it is not clear how the benefits of lock extensions can be evaluated adequately without first managing waterway traffic more efficiently on the existing system, (NRC, 2001, p. 4)." The committee further concluded that "the benefits and costs of lock extensions should not be calculated until nonstructural measures for waterway traffic management have been carefully assessed" (NRC, 2001, p.4).

A contingent congestion toll would internalize an important externality, providing incentives to low-value traffic to pass through the locks at less congested periods and to high-value traffic to minimize the time that others are made to wait. The extent to which investment in other nonstructural measures (e.g., "helper boats") likewise would help relieve congestion should also be evaluated.

The greatest benefit to the nation would be achieved by first implementing cost-effective, nonstructural alternatives and subsequently evaluating the benefits and costs of lock extensions once new traffic patterns have been established. Although implementing such measures may obviate the need for immediate lock extensions, they would not necessarily preclude eventual construction. Current laws, however, do not permit necessarily Corps to implement charges such as a congestion toll.

UMR-IWW Ecosystem

A thorough assessment of the ecological implications of the proposed lock extensions and additional towboat traffic will require a better understanding of the river's ecological system dynamics than currently exists. In particular, information regarding the effects of the existing system of locks, dams, navigation pools, and towboat traffic on river-floodplain ecology is essential. To improve scientific knowledge of Upper Mississippi River ecology, the Corps, the U.S. Geological Survey, and the U.S. Fish and Wildlife Service have been cooperating in the Environmental Management Program (EMP) on the Upper Mississippi River since 1986. The EMP is conducted through the Upper Midwest Environmental Science Center in La Crosse, Wisconsin, and is viewed by many as the nation's premier river monitoring program (see also *http://www.mvr.usace.army.mil/EMP/*; accessed September 16, 2003). The EMP has gathered and synthesized a large amount of scientific data and has issued many reports on river ecology (see, for example, USGS, 1999, 2000). Important gaps remain in scientific understanding of Mississippi and Illinois River ecology, however, and the Phase I committee recommended additional studies to help fill these gaps: "Gaps in current scientific understanding make it very difficult to accurately understand how additional changes will affect the river. . . . Systemwide research should be conducted on . . . cumulative effects of the existing navigation system on river ecology" (NRC, 2001, p. 6).

Despite the need for additional data and studies, the complexities of an ecosystem on the scale of the Upper Mississippi and Illinois Rivers must be recognized, as must the limited ability of scientific inquiry to reduce uncertainties in the understanding of ecosystem dynamics. Some degree of uncertainty will always be present in ecological knowledge of a system such as the UMR-IWW. Additional investigations are merited, but scientists and managers should not become caught up in a quest for certainty. At some point, scientists and managers must decide that existing data are sufficient to allow management actions to be implemented. Outcomes from management actions should be monitored, with findings being used to adjust future management actions. Recognizing uncertainties and the limits of knowledge, acting in the face of uncertainties, and monitoring management actions are tenets of an "adaptive management" approach that the Corps is discussing with regard to the UMR-IWW.

Integrated Waterway Systems Planning

The Upper Mississippi and Illinois Rivers and their floodplains support a variety of activities in addition to commercial navigation, including boating, recreational and commercial fishing, hunting, camping, and sightseeing. These activities are important to both the economics and the quality of life in dozens of communities along the rivers, including several cities. Studies of future UMR-IWW navigation thus require multidisciplinary perspectives. The Corps has conducted many river ecology studies, but it has been challenged to integrate them into the larger feasibility study. This prompted the Phase I committee to conclude, "The Corps should aim toward a more comprehensive and integrated assessment of the navigation system's effects on the environment in the UMR-IWW" (NRC, 2001, p. 4). Integration also would entail the consideration of social and cultural issues. Although some of these issues may go beyond the Corps' ability to address within its analytical framework, the Corps has conducted many meetings along the river in which it has discussed the feasibility study with members of the public.

Considerable effort has gone into broadening the scope of the study to identify and address not only the ecological impacts of the navigation improvements, but also the ecological impacts of the baseline operations of the navigation system. Alternatives are being developed to address these impacts. At the same time as this study is being carried out, the Corps (in collaboration with several other federal agencies) is developing an *Upper Mississippi River Comprehensive Plan.* This plan was authorized in Section 459 of the 1999 Water Resources Development Act (WRDA) and is being conducted to create a systematic, integrated strategy for managing flood risks, nutrients and sediment (including bank erosion), environmental stewardship, and river-related recreation needs and expectations on the Upper Mississippi River.

Although the Comprehensive Plan surely will use information generated by the restructured feasibility study, there is no indication that it will address the potential to deal concurrently with navigation and ecosystem restoration. For example, extensive levees in the lower section of the Upper Mississippi affect navigation pools and river-floodplain ecology. A navigation solution that does not consider restoring some connectivity between the river and the floodplains behind levees for environmental purposes will have missed an opportunity. Similarly, there appears to be little in the study dealing with opportunities to improve water quality of the Upper Mississippi River.

The Corps has adopted a more holistic approach to the feasibility study since the NRC (2001) Phase I report, and this committee commends its efforts to broaden the investigations. In reality, the Upper Mississippi River and Illinois Waterway region is one system, with human society interacting with natural systems. An integrated, or holistic, water management approach recognizes this and seeks to explain the relations between ecology, economics, and people, both quantitatively and qualitatively.

In a large, complex system with as many users and uses as the UMR-IWW, activities in different sectors often impinge on one another. For example, deepening of the navigation channel and expansion of the navigation system over past decades have had effects on other sectors and users, especially river ecology. The interface and the trade-offs between these two sectors are at the center of many differences of opinion about how the river should be managed and developed in the future. The divergent opinions that the committee heard in briefings by representatives of several interest groups at the September 2003 meetings indicate the difficulties that the Corps faces in managing this complicated system. Future decisions regarding UMR-IWW navigation system management are likely to have significant effects on river ecology, communities, and related human activities. The feasibility study should recognize the near inevitability of trade-offs between sectors and explain clearly how those trade-offs will be considered.

A note on the roles of the U.S. Congress in helping make decisions about trade-offs also is in order. With its many publicly owned locks and dams, wildlife refuges, and other lands, the UMR-IWW is both an interstate and a federal resource. Appropriate allocation of this system's benefits for navigation, environmental services and values, and flood control is thus ultimately a decision for Congress. Although the Corps of Engineers can help reduce uncertainties in the feasibility study, decisions regarding priorities for uses of this public resource are beyond the agency's purview. The Corps should conduct credible technical analyses, but when the agency must decide on trade-offs between different users, clear direction from Congress would be useful. In 1986, Congress passed an Upper Mississippi River Management Act (P.L. 99-662) that stated,

> It is hereby declared to be the intent of Congress to recognize that system as a nationally significant ecosystem and a national significant commercial navigation system. Congress further recognizes that the system provides a diversity of opportunities and experiences. The system shall be regulated and administered in recognition of its several purposes.

Although the act recognizes the multiple purposes of the river, it provides no guidance to the Corps on how to balance competing uses and inevitable trade-offs between sectors. Chapter 2 contains findings and recommendations and concludes this committee's first report to the Corps.

2

Findings and Recommendations

SPATIAL PRICE MODEL AND ESSENCE

The report from the Phase I committee stated that neither the September 2000 form of the spatial price model nor the results of the ESSENCE model should be used in the feasibility study (NRC, 2001). That review recommended structural changes in the ESSENCE model and the incorporation of empirically determined coefficients of the elasticity of waterway traffic demand. More generally, the Phase I committee report advocated the development of a suitable spatial price model. In response, the Corps stated that further development of the spatial price model should occur but "in a research and development setting outside the study process" (USACE, 2003b, p. 1).

This committee did not explore the details of the Tow Cost Model; however, the TCM is not widely accepted by economics experts, or even by the Corps, as a useful tool for modeling water transportation demand for grain. Furthermore, the decision to adopt the Tow Cost Model contradicts advice about demand modeling provided in the NRC Phase I report by an independent group asked for advice in improving the study's economic analyses.

The Corps proposes that national economic development (NED) benefits from navigation improvements will be estimated from the Tow

Cost Model and, alternatively, from the ESSENCE model, using two different but arbitrary values for the *N* coefficient (elasticity of demand for waterway transport). No supporting data have been presented to indicate why these are in fact lower and upper bounds or what might be the best estimates of *N*. Furthermore, the ESSENCE model, in its original form, adopts a functional form that should be regarded as a highly simplified approximation. When used outside a narrow range, it yields implausible results. The nature of the current application across time and space stretches that simplified specification beyond the range in which it might be a useful approximation. Few of the Phase I committee's key recommendations for enhancing the value and credibility of the ESSENCE model have been implemented, and this committee finds the absence of a spatial price model unacceptable. **There is no useful role for the ESSENCE model in the restructured feasibility study.**

The Corps is to be commended for initiating the development of a spatial price model. This committee, however, finds that the steps taken in the restructured feasibility study represent inadequate responses to the NRC Phase I report. Model development efforts have not adopted, for example, realistic assumptions regarding spatial variation in grain production and shipping costs, the range of ports that might be accessed by regional grain producers, domestic processing demands and the location of these demands, or global grain supplies and demands. The restructured study also assumes that the division of grain exports among available ports will not change, which is an unlikely assumption. As lock congestion builds on the U.S. inland waterway system, domestic markets and alternative ports and routings become increasingly feasible and likely. For example, overland grain exports to Mexico may increase as lock congestion builds, as may rail shipments to Pacific Northwest ports or shipments to domestic markets now served by Mississippi River basin grain production. Moreover, since 80 percent of U.S. corn production is consumed domestically, some dimension of this demand should be explicitly modeled. With some improvements and adjustments, existing spatial grain models could be adapted to give superior insight to the approaches currently considered by the Corps. Appendix A identifies and discusses key factors that should be included in the development of a credible spatial price model(s). Appendix A also describes a transportation demand model developed for the Panama Canal (see Fuller et al., 1999). This committee has not sufficiently studied the Panama Canal transportation demand model to be able to recommend it specifically for use in the UMR-IWW study; however, it is a fully developed model that goes a long way toward incorporating the elements of a full spatial equilibrium model, and it merits investigation by the Corps.

If the Corps develops its own spatial price model, this development could proceed in clearly defined modules. One module should forecast the amount of grain grown in the upper Midwest, which will be a function of the cost of growing grain and other commodities compared to prices at which grains and alternative commodities could be sold. Another module should examine grain production in other grain-producing regions around the world (especially Argentina and Brazil) and associated prices. Another module should focus on world demand for grain, which is a function of population, income, domestic production, and global market prices of meat for import. Appendix A lists additional factors that should be included in a credible spatial price model of UMR-IWW transportation demand. If these types of improvements and adjustments are to be made and incorporated into the feasibility study, the current schedule for study completion will have to be relaxed.

Demand Forecasts

The Corps should be commended for applying a scenario approach to forecasting future waterway traffic demand. The Corps contracted with Sparks Companies, Inc., of Memphis, Tennessee, to provide a report depicting a range of future economic scenarios and resulting demand for barge transportation (Sparks Companies, 2002). Drawing from the Sparks study, the Corps is using five scenarios of future grain exports. Four of the five scenarios assume substantial increases in exports, and the fifth assumes little to no change in current levels.

After exhibiting increasing trends from 1950 to 1980, U.S. grain exports have since shown almost no growth (USDA, various dates). There is little compelling evidence for a substantial increase in world demand for U.S. grain exports over the next decade or two. Indeed, several factors could contribute to declining grain exports. For example, there is an increasing tendency among Asian nations to import U.S. meat products directly, rather than to import grain for feed in their own nations. Further, South American nations such as Brazil are increasing domestic soybean production levels, which could reduce global grain market demands for U.S. exports. Yet despite nearly 25 years of essentially stable U.S. grain export levels, four of the five scenarios in the feasibility study assume substantial increases in export levels. Given the relatively flat level of exports over the past two decades, the committee views the projected increases in four of five scenarios with some skepticism. **Forecasts of increases in U.S. grain exports should present explanations for likely export trends after 2003 that are consistent with history and with expert opinion on likely future conditions in global grain markets.** The committee looks forward to discussing the assumptions, methods, and projections with the Corps' consultant at our next meeting.

MANAGING WATERWAY CONGESTION

The Phase I committee report urged the Corps to conduct a "comprehensive review and assessment of nonstructural options for improving traffic management" (NRC, 2001). The Corps then requested the Volpe National Transportation Systems Center to evaluate a number of traffic management measures. In addition, the Corps conducted its own analysis of congestion fees (to be implemented through a lockage fee). This committee commends the Corps for seriously considering a lock usage fee, even though existing legislation prohibits implementation of such a scheme. Economic incentives, including lock usage fees, should produce net economic benefits. However, a true congestion fee is a *contingent fee* that is levied only at times of congestion and only on tows that contribute to congestion. Although implementing such a fee is admittedly not a simple matter, the resulting benefits should be greater than those obtained from a simple lock usage fee of comparable magnitude.

At the committee's September 2003 briefings, it was stated that other traffic scheduling alternatives had been analyzed by the Volpe National Transportation Systems Center and were found infeasible. The report from the Volpe group was not available to this committee at its September 2003 meeting, and details were not provided at that briefing. The committee subsequently received the Volpe report (Dyer et al., 2003), which it looks forwarding to discussing with the authors. In addition to a congestion fee, the Corps should evaluate other nonstructural measures for improving traffic management, such as tradable arrival slots and industry self-help systems.

The Corps' feasibility study maintains that the "without-project" condition will include continued use only of the current traffic management system (mainly a "first come, first served" system). This is contrary to findings of the 2001 NRC report, which stated that the benefits of proposed lock extensions to the existing system cannot be evaluated fully until the existing system is operated more efficiently and recommended that the Corps apply a wider range of options for managing congestion. Like the report from the Phase I committee, this report also finds that meaningful planning of lock extensions must await the time when the existing system is operated at reasonably full efficiency and that the without-project condition should include traffic management measures that achieve more effective operational efficiency of the existing system. Moreover, because improved waterway traffic management should shorten the idle time of tows at locks and between locks, such measures could also reduce environmental impacts such as fish entrainment and increased turbidity.

This committee appreciates the challenges of implementing such measures. Developing and implementing an effective nonstructural traffic management system(s) on the UMR-IWW will not be simple, quick, or inexpensive, and it will clearly entail both progress and setbacks and will require time and resources. The time required to implement and evaluate a nonstructural scheme(s) suggests that a relaxation of the current feasibility study schedule is in order. Technologies for establishing such a system are available and hold promise for reducing the costs of congestion. **The Corps should proceed as soon as practicable toward developing and implementing a nonstructural system to help alleviate waterway traffic congestion.**

INTEGRATED SYSTEMS PLANNING

River Ecology

The initial feasibility study for the UMR-IWW had a narrow focus with regard to environmental concerns and issues of environmental sustainability. The restructured study is assuming a broader focus on these issues, and the committee commends the Corps for broadening the scope to deal not only with the ecological impacts of lock extensions and other measures for enhancing navigation, but also with the ecological impacts of baseline operations of the existing system of locks and dams and navigation pools. Indeed, the Corps' Interim Report (USACE, 2002, p. 18) acknowledges that economic and ecological needs should be of equal priority. A broad, holistic perspective is also necessary because of the significant implications of Mississippi River water quality and sediment transport for downstream regions in and along the Gulf of Mexico. The Corps should thus, to the maximum extent feasible, consider factors such as water quality, flood damage reduction, and sediment transport in order to reflect a more holistic approach to dealing with the diverse management issues in the UMR-IWW.

The Interim Report (USACE, 2002) describes many management alternatives that the Corps is considering not only to mitigate effects of proposed navigation system expansion, but also to repair environmental damages caused by the existing system of locks and dams and navigation pools. Nonetheless, there may be problems in achieving the environmental goals because as that report noted, "[E]cosystem restoration is not a specifically authorized purpose of the 9-Foot Channel Navigation Project" (USACE, 2002). Congressional action may thus be necessary and appropriate for expanding the purposes of the UMR-IWW to include environmental sustainability.

A general observation is that the Corps' analysis regarding river ecology is less advanced than its navigation studies, due largely to the relatively recent broadening of the feasibility study to better evaluate and integrate ecological concerns. The committee's comments on these aspects of the study are similarly less specific than its comments on the economics of waterway transportation.

Adaptive Management

The recommendations from the Phase I committee included a proposal that the Corps implement adaptive management principles within the feasibility study. The Phase I committee was concerned with the "adaptive mitigation" strategy discussed in the feasibility study, which the committee found "inconsistent with the principles of adaptive management articulated in the natural resources management literature" (NRC, 2001, p. 7). In responding to the NRC (2001) report, the Corps' background summary (USACE, 2003a) stated that the "mitigation plan will incorporate the principles of adaptive management." This committee regards that assurance as a positive sign but awaits further evidence of progress.

Adaptive management does not apply solely to mitigation, as some sections of Corps reports imply. The approach is equally valuable in planning ecosystem restoration activities (see following section) and in planning and implementing structural and nonstructural solutions to waterway congestion. Furthermore, adaptive approaches should aid in developing strategies for large, unanticipated events such as floods. This is especially important in the UMR-IWW context because floods have long posed management challenges along the UMR-IWW. Flood problems could also be affected by regional changes in climate patterns. An adaptive management approach should help the Corps better understand how to adjust to future changes, and the Corps should consider the possibility of changes such as long-term changes in climate. The adaptive management approach can also help focus attention on nonstructural solutions that avoid, in the words of the Phase I committee, "the trap of irreversibility."

Many of the environmental restoration, mitigation, and environmental enhancement actions proposed for the Upper Mississippi reflect a collaborative effort between the Corps and the U.S. Fish and Wildlife Service. Many of the ongoing ecological investigations reflect current techniques and current ecological understanding. Although the approaches are understood, the probabilities regarding implementation and resulting ecological outcomes are uncertain. Moreover, the Corps has had only marginal success in gaining support from the administration and Congress for ecological projects of such magnitude. Given these conditions, it is appropriate that this effort be conducted within an adaptive management framework, Although this framework is described in the Corps' 2002 Interim Report only in the context of mitigation, it features prominently in the draft 2003 report (Lubinski and Barko, 2003) from an environmental science panel formed by the Corps as part of the feasibility study and espoused by the authors as "the overarching theme of future integrated efforts for management of the UMR-IWW" (p. 1). Such an adaptive approach will involve carefully and openly crafting experiments, closely monitoring results, and consulting both stakeholders and program objectives to adjust plans accordingly. It will also require the involvement of stakeholders, an organizational structure to oversee the efforts, and a fiscal commitment to support the activities.

The Corps' Interim Report and the 2003 draft science panel report describe some processes that might be amenable to adaptive management (e.g., the Corps has undertaken some limited experimentation in the drawdown of Pool 8 to improve ecological conditions), but they also note potential difficulties. Given the extent of this effort and the probable need for adaptation in progress, the adaptive management paradigm would nonetheless seem ideal. Recent National Research Council reports on the Missouri River ecosystem (NRC, 2002) and on Glen Canyon Dam operations (NRC, 1999) describe the use of such adaptive management processes in more detail. **The Corps should implement adaptive management concepts and approaches throughout all aspects of the planning process.**

Trade-offs

On a river system used as intensively as the UMR-IWW, enjoyment of the system's benefits by one user or sector typically has effects on other users or other sectors. For example, increases in towboat traffic will have negative effects on fish and river ecology; the drawdown of navigation pools is good for river ecology, but these actions can expose sandbars and mudflats, which can be bad for boaters. As a result, management decisions that enhance benefits for one sector without diminishing benefits for another (so called win-win scenarios) are likely to be small in number and extent in comparison with more contentious scenarios that involve trade-offs—with winners and losers—between users and sectors. Many of these latter types of scenarios are underpinned by fundamental differences in values and perceptions, yet traditional planning approaches offer little more than analytical results for solution. Better guidance from Congress and the administration on how to prioritize river uses and to weigh major trade-offs would be useful to the Corps, which is limited in its ability to allocate federal resources among competing objectives and users.

Even in the absence of new guidance, some things can be done to facilitate trade-offs between users of the benefits from the Mississippi and Illinois Rivers. Some aspects of environmental restoration are amenable to quantification and monetization, such as instances in which restoration obviates the necessity of costly mitigation or where the public may exhibit measurable willingness to pay for the improvement. In the latter case, revealed preference methods, such as the travel cost method or hedonic price analysis, may be employed in specific instances. Stated preference methods, such as contingent valuation analysis, can be applied in most if not all instances. There is a rich literature on these topics, with hundreds of published texts and papers, including some published by the Corps. Although these approaches may be partial in coverage or subject to some error, there are instances in which even partial information is sufficient to resolve a trade-off. However, the committee is not aware of any environmental valuation studies being performed in conjunction with the feasibility study, so the Corps may be left with only qualitative approaches at its disposal.

DECISIONS, IMPLEMENTATION, AND INSTITUTIONS

Timing

The Corps is on an aggressive timetable for finishing its feasibility study, especially in regard to its ability to consider this committee's recommendations, the bulk of which may be issued only months before a "Chief's Report" is scheduled (November 2004). Not only does the current schedule provide the Corps with a short amount of time to respond to the present and subsequent reports from this committee in general, but implementation of this report's recommendations cannot be adequately completed on the current feasibility study schedule. Although the committee is prepared to conduct its review and provide advice to the Corps with due speed, and likewise respects the need to move forward with the feasibility study, the study schedule should allow adequate time for credible analysis to be conducted and concluded. **The Corps should extend its schedule for completing the feasibility study and issuing a Chief's Report.**

Prioritization and Sequencing

In the September 2003 discussions with this committee, Corps staff members outlined progress on an environmental restoration plan, developed in collaboration with federal and state partners. This progress is commendable and addresses the Phase I committee report recommendation that the Corps recognize that "environmental concerns have become a core issue in the operation of inland waterways systems" and therefore should "adapt its planning, engineering design, operations, and analysis accordingly" (NRC, 2001).

During the September 2003 briefing, the Corps presented a map of a representative navigation reach that displayed the proposed restoration projects developed by the Corps and its partners. Virtually every square meter of the reach was identified as needing some type of rehabilitation, and it became obvious that some criteria and a process will be essential for prioritizing and sequencing these projects. One such criterion might be to give priority to projects that would restore natural processes, with the expectation of triggering self-repair and self-maintenance over large areas at relatively modest cost. For example, operating dams to restore a more natural flood pattern (spring flood followed by stable, low water levels during the summer growing season) would build on successes already achieved in pool drawdown experiments. The compaction and drying of sediments, and the recovery of vegetation itself, then would help stabilize the banks and bottoms of backwaters, and the plants might fulfill their functions of taking up nutrients and providing food and shelter for wildlife. Priority might also be given to projects that meet multiple objectives, including (as suggested above) flood damage reduction. For example, flood easements or outright buyouts of selected levee districts might reduce future flood damages and flood heights, reduce the time during floods when the river has to be closed to navigation, and provide fish and wildlife habitat. A fundamental principle of adaptive management is evaluating the outcomes of management actions (evaluation builds upon and draws lessons from data gathered from monitoring programs). Since the Corps intends to implement the feasibility study with an adaptive management approach, outcomes of early restoration projects should be evaluated carefully. The results of these evaluations will be useful in determining the costs and benefits associated with different restoration approaches. **Priority should be given to restoration projects that aim to restore natural processes and those that aim to achieve multiple objectives.**

The sequencing of projects should also be considered. For example, restoration of more natural hydrology throughout an entire navigation pool might restore vegetation in some areas without restoration projects or might require less expenditure on levees and pumps to promote favorable water levels within constructed floodplain impoundments. In this case, it might be useful to modify dam and pool operations before constructing or renovating wildlife management areas within the floodplain.

Cost Sharing and Funding

In addition to prioritization and sequencing, there are important issues related to funding and cost sharing. Several speakers at the committee's September 2003 meeting cited a previous history of Corps, administration, and congressional failure to concurrently fund navigation improvements and ecological protection and recovery. The general concern was that navigation projects are completed but the related environmental projects are funded at low levels or not at all. The best environmental plan (under the new dual-track planning process) will be of little use if it is not implemented. Suggestions for securing funding for ecosystem restoration in the UMR-IWW include a program similar to the Land and Water Conservation Fund or a trust fund similar to the Inland Waterways Trust Fund. The Corps has identified the potentially high costs of implementing proposed restoration and adaptive management efforts as part of the feasibility study. In order to maintain the study's credibility and ensure integration across sectors, it is important that efforts to enhance environmental benefits be carried out concurrently with efforts to improve navigation.

There is less federal land along the Mississippi River in the downstream states (e.g., Missouri and Illinois) than in the upper basin. Floodplains in Iowa, Illinois, and Missouri are mainly privately held, whereas much of the

Mississippi River floodplain in Minnesota and Wisconsin is part of the U.S. federal Upper Mississippi River Fish and Wildlife Refuge system. On federal lands, the federal share of restoration projects is 100 percent; on private lands, however, local sponsors are responsible for part of the costs. In the basin's downstream states, most of the restoration projects will thus require local sponsors to provide 35 percent of the planning and construction costs and all of the long-term maintenance. Some observers have called for changes in cost-sharing arrangements to ensure that worthy restoration projects in the downstream states receive similar consideration as restoration projects in the upper basin.

Construction costs for extension of the locks and associated environmental mitigation are funded with 50 percent of the construction and environmental mitigation costs coming from the Inland Waterway Trust Fund and the remaining 50 percent from congressional appropriations. In contrast, environmental restoration components of the project generally require non-federal cost sharing:

- Environmental Management Program projects as authorized by the Water Resources Development Act of 1986 and amended in the WRDA of 1990 and 1999 provide 100 percent federal funding for EMP projects on National Wildlife Refuge lands and 65 percent federal, 35 percent non-federal funding for projects not located on federal lands.
- For modification of structures and operations of water resources projects to improve the quality of the environment under Section 1135 of the WRDA of 1986, the cost sharing is 75 percent federal, 25 percent non-federal.
- For aquatic ecosystem restoration and protection projects developed under Section 206 of the WRDA of 1996, the cost sharing is 65 percent federal, 35 percent non-federal.
- For projects under Section 204 of the WRDA of 1992 for the protection, restoration, and creation of aquatic and ecologically related habitat in conjunction with the dredging of authorized navigation projects, incremental costs of the beneficial use of dredged material for habitat creation are shared—75 percent of the costs are borne by the federal government, and 25 percent are borne by a non-federal sponsor.

Furthermore, there may be limited or no federal funding available to implement nonstructural measures because the Corps lacks the necessary authority to implement and/or fund such measures. This is an analytical concern because if proposed restoration projects are not implemented, the results from analyses assuming that these projects will be implemented (and will deliver benefits) will be discredited in proportion to the extent that the projects are not funded. **The Corps should identify instances in which federal cost-sharing rules are likely to restrict or preclude implementation of environmental restoration projects and nonstructural measures.**

ENGINEERING

Construction Cost Contingency

The Phase I committee report (NRC, 2001) recommended that the Corps increase its standard contingency factor from 20 to 25 percent because of the uncertainty associated with cost estimates for lock extensions. In its background summary response (USACE, 2003b), the Corps stated: "Jacobs Engineering, an independent engineering firm, reviewed representative large-scale cost estimates developed in the original study. The review validated that the estimates developed for lock construction alternatives are reasonable." In response to this committee's September 2003 request for the Jacobs Engineering report, the Corps indicated that the report was still being drafted. It will be necessary to obtain and review this report before this committee can comment on the Corps response.

Lock and Dam Rehabilitation

The NRC (2001) Phase I committee report recommended: "If new waterway traffic demand forecasts are developed, it will be important to revisit the rehabilitation costs analysis to ensure consistency with the revised traffic demand forecasts." The Corps background summary response (USACE, 2003b) does not reply to this recommendation. This committee requests that the Corps respond to this recommendation from the NRC 2001 report.

References

Dyer, M. G., P. K. Zebe, A. Rao, and M. C. Caputo. 2003. Draft Upper Mississippi River and Illinois Waterways: Non-Structural Measures Cost-Benefit Study. Cambridge, MA: U.S. Department of Transportation, Research and Special Programs Administration.

Fellin, L., and S. Fuller. 1997. Effect of the proposed waterway user tax on U.S. grain flow patterns and producers. Journal of the Transportation Research Forum 36:11-25.

Fuller, S., L. Fellin, and W. Grant. 1999. Grain transportation capacity of the Upper Mississippi and Illinois Rivers: A spatial analysis. Journal of the Transportation Research Forum 38:38-54.

Koo, W. 1985. Tariffs and transportation costs and U.S. wheat exports: A quadratic programming model. In Transportation Models for Agricultural Products, W. Koo and D. L. Larson, eds. Boulder, CO: Westview Press.

Lubinski and Barko. 2003. Upper Mississippi River-Illinois Waterway System Navigation Feasibility Study: Environmental Science Panel Report. Draft report. Rock Island, IL, St. Louis, MO, and St. Paul, MN: U.S. Army Corps of Engineers.

Martin, L. 1981. Quadratic single and multiple-commodity models of spatial equilibrium: A simplified exposition. Canadian Journal of Agricultural Economics 29:21-48.

National Research Council (NRC). 1999. Downstream: Adaptive Management of Glen Canyon Dam and the Colorado River Ecosystem. Washington, DC: National Academy Press.

National Research Council. 2001. Inland Navigation System Planning: The Upper Mississippi River-Illinois Waterway. Washington, DC: National Academy Press.

National Research Council. 2002. The Missouri River Ecosystem: Exploring the Prospects for Recovery. Washington, DC: National Academy Press.

Sparks Companies, Inc. 2002. Upper Mississippi River and Illinois Waterway Navigation Study. Economic Scenarios and Resulting Demand for Barge Transportation. Final Report. Memphis, TN.

U.S. Army Corps of Engineers (USACE). 2002. Interim Report for the Restructured Upper Mississippi River-Illinois Waterway System Navigation Feasibility Study. Rock Island, IL: Rock Island District Corps of Engineers.

U.S. Army Corps of Engineers. 2003a. Upper Mississippi River-Illinois Waterway System Navigation Study: Responses to February 2001 NRC Review.

U.S. Army Corps of Engineers. 2003b. Monthly Status Report. Upper Mississippi River-Illinois Water System Navigation Study. Rock Island, IL: Rock Island District Corps of Engineers.

U.S. Department of Agriculture (USDA). Various dates. Agricultural Statistics. National Agricultural Statistics Service. Washington, DC: U.S. Government Printing Office.

U.S. Geological Survey (USGS). 1999. Ecological Status and Trends of the Upper Mississippi River System 1998. *http://www.umesc.usgs.gov/reports_-publications/status_and_trends.html*.

U.S. Geological Survey. 2000. Upper Mississippi Environmental Management Program: Habitat Needs Assessment. *http://www.umesc.usgs.gov/-habitat_needs_assessment/*.

Appendix A

Spatial Equilibrium Models and the U.S. Grain Sector

The ESSENCE Model

The amount of grain that will be transported by towboats on the Upper Mississippi River-Illinois Waterway (UMR-IWW) in future years depends on many factors, including the following:

- the amount of grain grown in the river basin area, which depends on the cost of growing grains and alternative commodities compared to the price at which grains and alternative commodities can be sold, since the land could be used for other crops or left fallow;
- the amount of grain grown in other areas of the world, particularly Brazil and Argentina, and the price at which it is offered for export;
- the world demand for grain, particularly the demand for imported grain in each nation, which depends on population, income, home agricultural output, tastes, and the world price of meat for importation;
- the price of transporting grain from each exporting port to each importing port;
- the price of transporting grain from each growing area to each exporting port;

- the domestic demand for grain, including raising animals for export;
- the current logistic and transportation prices to transport U.S. grain from multiple origins to relevant domestic markets;
- prices of grain at relevant export and domestic markets; and
- the use of current railroad shipping rates, rather than rates from old railroad tariffs or estimated costs from rail cost models.

Only after determining the effects of these (and perhaps other) factors can future demand for grain transport on the UMR-IWW be credibly estimated.

The Phase I committee recommended that the ESSENCE model not be used in the feasibility study. This model includes a traffic routing component and a simple reduced form economic model for estimating the effect of shipping cost on barge movements. ESSENCE is not a spatial equilibrium model. It does not allow for alternative export ports or alternative destinations for the grain. It is silent on most of the factors noted above that are part of a full spatial equilibrium model. Furthermore, the portion of the model used to determine the impact of shipping cost is seriously misspecified.

This model should not be criticized because it is simple since simple models can be useful if they describe historical experience well and if their implications are plausible. The Phase I committee noted that this model was never fit to historical data, so this simplification cannot be justified on the basis of being a good description of current data. Furthermore, the shipping demand curve used in ESSENCE is defined by a single parameter N. The implications of the model are implausible when N is outside a narrow range. For example, N is characterized by the Corps as describing the price elasticity of barge rates. This is not correct. Price elasticity is not constant for a fixed value of N.

The relationships embodied in ESSENCE would not fit historical data well because they do not account for the level of foreign demand, competition with foreign grain, U.S. domestic demand, or other important factors. The Corps should thus abandon this model, including any reference to values of N as lower or upper bounds. The Corps should not to spend time and effort fitting this model to historical data. Even if the model fit historical data well, it would not serve as a building block in implementing a full spatial equilibrium model.

Spatial Equilibrium Models

The rationale underpinning the spatial equilibrium concept was stated in 1951 as follows: two or more regions with demand and supply functions produce and consume a homogeneous product and are separated by a known transfer cost. Given this information, the problem is to determine the equilibrium levels of production, consumption, and prices in each region and the equilibrium trade flows between regions (Enke, 1951). A geometric interpretation of this problem and its solution for two regions was later developed (Samuelson, 1956), and Takayama and Judge (1971) showed how the problem could be formulated as a mathematical programming problem involving any number of regions.

Numerous spatial equilibrium models of the U.S. crop sector have been developed to address issues relating to interregional trade and agricultural policy, while others have featured considerable transportation system detail for purposes of evaluating transportation policy and infrastructure issues. These models feature regional demand and supply relationships and linking transfer costs and are capable of examining the effect of transportation infrastructure improvements or policies on commodity prices, production, trade, and welfare measures.

A study (Fuller et al., 2000) that employed spatial price equilibrium models to determine the effect of Panama Canal closure and alternative toll levels on U.S. agriculture may be of interest to the Corps as it proceeds with the UMR-IWW feasibility study. This analysis of Panama Canal transportation demand showed that increases in toll rates would reduce grain exports via the U.S. Gulf of Mexico ports, reduce grain flow on the Upper Mississippi and Illinois Rivers and lower regional grain prices, increase exports via U.S. Pacific Northwest ports and rail movements to these ports, reduce quantities transiting the Panama Canal, and increase maritime movements to East Asia via Africa's Cape of Good Hope. Further, the U.S. role in the Asian corn and soybean markets would decline, while competitors' roles (Argentina, China) would increase. Analyses indicated that the Panama Canal had a value to U.S. corn and soybean producers of about $300 million per year. (These models are executed on desktop computers with 256 MB

RAM and CPU of 1.3 GHz [Pentium 4] in about 20 minutes.) Although the model is not a perfect analogue for evaluating all issues related to lock congestion on the Upper Mississippi and Illinois Rivers, it is a fully developed model that incorporates many elements of a full spatial equilibrium model, and it merits investigation by the Corps for its relevance to the UMR-IWW feasibility study.

Existing corn and soybean market models are spatial, intertemporal equilibrium models that include the domestic and international sectors (see Fellin and Fuller, 1997; Fuller et al., 1999; Koo, 1985; and Martin, 1981). These models include details on regional excess demands and supplies and on transportation, storage, and grain handling costs in the United States. Other trading countries, excluding Mexico, are treated as either an excess supply or an excess demand region. Mexico includes excess demand and supply regions, and linking transportation costs with grain handling and storage costs.

The international corn model includes 78 excess supply regions and 99 excess demand regions. The excess corn demand regions include 64 U.S. regions, 8 Mexican regions, and 6 foreign regions. Included among the excess corn demand regions are 60 U.S. regions, 14 Mexican regions, and 24 foreign demand regions. With the exception of Japan, South Korea, China, Canada, and Taiwan, the foreign excess demand regions are an aggregation of countries. The international soybean model includes 80 excess supply regions and 58 excess demand regions. Of the excess supply regions, 68 are located in the United States, 8 in Mexico, and 4 foreign supply regions that represent Argentina, Brazil, Paraguay, and Bolivia. The excess soybean demand regions include 24 U.S. regions, 9 Mexican regions, and 24 foreign excess demand regions.

Embedded in the United States and Mexico portions of the models are extensive transportation networks that connect excess supply regions with excess demand regions and ports via truck, rail, and barge costs or rates. Excess supply regions are linked by truck and rail to 37 barge-loading facilities on the inland waterway system. The barge loading sites are linked to barge unloading sites on the U.S. inland waterway system and ports as appropriate. Of the U.S. ports, 17 receive corn and soybeans from excess supply regions via truck, rail, and barge as appropriate and then ship via maritime costs or rates to a representative port in each of the 24 foreign excess demand regions.

To represent winter freezing of the Great Lakes and Upper Mississippi waterways, the models disallow shipping via these arteries in the winter quarter. The models include four quarters. An excess supply region is represented by an inverse excess supply equation, while an excess demand region is represented by four (quarterly) inverse demands. Grain is produced in the fall quarter and carried through the subsequent crop year by the cost of storage. Excess supply regions are typically crop-reporting districts that include from 10 to 15 counties. As an example, in Iowa there are eight excess supply regions and one excess demand region. The typical excess supply region in Iowa is linked to 3 barge-loading sites (2 on the Mississippi River and 1 on the Missouri River); 28 domestic excess demand regions; 8 port areas (Chicago, Duluth, Mobile, New Orleans, Galveston, San Diego, Portland, and Seattle); and a U.S.-Mexico border-crossing site (Laredo) by truck and rail costs as appropriate. In addition, the model's barge-loading sites are linked by barge costs or rates to inland barge-unloading sites and selected U.S. port areas. Barge rates reflect seasonality, but from St. Paul, Peoria, and St. Louis to the Lower Mississippi River, ports average about $9.50, $7.30, and $5.40 per ton. U.S. ports are linked to 24 foreign excess demand regions by ship rates, and U.S.-Mexico border crossing sites are linked to 14 Mexican excess demand regions by Mexican rail rates. Further, foreign excess supply regions are linked to the 24 foreign excess demand regions by ship rates.

REFERENCES

Enke, S. 1951. Equilibrium among spatially separated markets: Solutions by electric analogue. Econometrica 19:40-48.

Fellin, L., and S. Fuller. 1997. Effect of the proposed waterway user tax on U.S. grain flow patterns and producers. Journal of the Transportation Research Forum, 36:11-25.

Fuller, S., L. Fellin, and W. Grant. 1999. Grain transportation capacity of the Upper Mississippi and Illinois Rivers: A spatial analysis. Journal of the Transportation Research Forum 38:38-54.

Fuller, S., L. Fellin, and K. Eriksen. 2000. Panama Canal: How critical to U.S. grain exports? Agribusiness: An International Journal 16(4):435-455.

Koo, W. 1985. Tariffs and Transportation Costs and U.S. Wheat Exports: A Quadratic Programming Model in Transportation Models for Agricultural Products, W. Koo and D. L. Larson, eds. Boulder, CO: Westview Press.

Martin, L. 1981. Quadratic single and multiple-commodity models of spatial equilibrium: A simplified exposition. Canadian Journal of Agricultural Economics 29:21-48.

Samuelson, P. 1956. Spatial price equilibrium and linear programming. American Economic Review 38:496-509.

Takayama, T., and G. Judge. 1971. Spatial and Temporal Price and Allocation Models. Amsterdam: North Holland Publishing Company.

Appendix B

National Research Council Board Membership and Staff

WATER SCIENCE AND TECHNOLOGY BOARD

RICHARD G. LUTHY, *Chair,* Stanford University, Stanford, California
JOAN B. ROSE, *Vice Chair,* Michigan State University, East Lansing
RICHELLE M. ALLEN-KING, University at Buffalo, Buffalo, New York
GREGORY B. BAECHER, University of Maryland, College Park
KENNETH R. BRADBURY, Wisconsin Geological and Natural History Survey, Madison
JAMES CROOK, Water Reuse Consultant, Norwell, Massachusetts
EFI FOUFOULA-GEORGIOU, University of Minnesota, Minneapolis
PETER GLEICK, Pacific Institute for Studies in Development, Environment, and Security, Oakland, California
JOHN LETEY, JR., University of California, Riverside
CHRISTINE L. MOE, Emory University, Atlanta, Georgia
ROBERT PERCIASEPE, National Audubon Society, Washington, D.C.
JERALD L. SCHNOOR, University of Iowa, Iowa City
LEONARD SHABMAN, Virginia Polytechnic Institute and State University, Blacksburg
R. RHODES TRUSSELL, Trussell Technologies, Inc., Pasadena, California
KARL K. TUREKIAN, Yale University, New Haven, Connecticut

HAME M. WATT, Independent Consultant, Washington, D.C.
JAMES L. WESCOAT, JR., University of Illinois at Urbana-Champaign

NRC Staff

STEPHEN D. PARKER, Director
LAURA J. EHLERS, Senior Staff Officer
JEFFREY W. JACOBS, Senior Staff Officer
WILLIAM S. LOGAN, Senior Staff Officer
LAUREN E. ALEXANDER, Staff Officer
MARK C. GIBSON, Staff Officer
STEPHANIE E. JOHNSON, Staff Officer
M. JEANNE AQUILINO, Administrative Associate
ELLEN A. DE GUZMAN, Research Associate
PATRICIA JONES KERSHAW, Study/Research Associate
ANITA A. HALL, Administrative Assistant
JON Q. SANDERS, Senior Project Assistant
DOROTHY K. WEIR, Project Assistant

TRANSPORTATION RESEARCH BOARD EXECUTIVE COMMITTEE

Chair: Genevieve Giuliano, Director, University of Southern California, Los Angeles
Vice Chair: Michael S. Townes, President and CEO, Hampton Roads Transit
Executive Director: Robert E. Skinner, Jr., Transportation Research Board

Michael W. Behrens, Executive, Texas Department of Transportation
Joseph H. Boardman, Commissioner, New York State Department of Transportation
Sarah C. Campbell, President, TransManagement Inc.
Joanne F. Casey, President and CEO, Intermodal Association of North America
James C. Codell III, Secretary, Kentucky Transportation Cabinet
John L. Craig, Director, State Engineer, Nebraska Department of Roads
Bernard S. Groseclose, Jr., President and CEO, South Carolina State Ports Authority
Susan Hanson, Professor of Geography and Director, Clark University
Lester A. Hoel, L.A. Lacy Distinguished Professor, University of Virginia
Henry L. Hungerbeeler, Director, Missouri Department of Transportation
Adib K. Kanafani, Cahill Professor of Engineering, University of California, Berkeley
Ronald F. Kirby, Director, Transportation Planning, Metropolitan Washington Council of Governments
Herbert S. Levinson, Principal, Herbert S. Levinson Transportation Consultant
Michael D. Meyer, Professor, Georgia Institute of Technology
Jeff P. Morales, Director, California Department of Transportation
Kam K. Movassaghi, Secretary, Louisiana Department of Transportation and Development
Carol A. Murray, Commissioner, New Hampshire Department of Transportation
David Z. Plavin, President, Airports Council International of North America

John H. Rebensdorf, Vice President, Network and Service Planning, Union Pacific Railroad Company
Catherine L. Ross, Director and Harry West Chair of City and Regional Planning, Georgia Institute of Technology
Paul P. Skoutelas, CEO, Port Authority Allegheny County
Michael W. Wickham, Chairman, Roadway Corporation

Appendix C

Committee Member and Staff Biographies

COMMITTEE MEMBERS

JOHN J. BOLAND (*chair*) is a professor in the Department of Geography and Environmental Engineering at Johns Hopkins University. His fields of research include water and energy resources, environmental economics, and public utility management. Dr. Boland has studied resource problems in more than 20 countries and has published more than 125 papers and reports, and coauthored two books on water demand management, and two others on environmental management issues. Dr. Boland is a registered professional engineer. He has served on several National Research Council (NRC) committees and panels, including the Water Science and Technology Board, of which he was a founding member (1982) and second chair (1985-1988). He is a life member of the American Water Works Association and past chairman of its Economic Research Committee. Dr. Boland received his Ph.D. degree in environmental economics from Johns Hopkins University.

PATRICK BREZONIK is a professor in the Department of Civil Engineering and the past director of the Water Resources Center at the University of Minnesota. Prior to his appointment at the University of Minnesota, Dr. Brezonik was a professor of water chemistry and environmental science at the University of Florida. His fields of research include biogeochemical processes in aquatic systems, with an emphasis on the impacts of human activity on water quality and element cycles in lakes and watersheds. He has served as a member of the National Research Council's Water Science and Technology Board and as a member of several NRC committees, including chair of the Committee to Revitalize Education in the Field of Limnology. He received his B.S. degree in chemistry from Marquette University and his M.S. and Ph.D. degrees in water chemistry from the University of Wisconsin-Madison.

ROBERT K. DAVIS has most recently been associated with the Institute of Behavioral Science at the University of Colorado. He is the former head of the Economic Staff in the Office of the Secretary of the U.S. Department of the Interior. His fields of research include natural resource economics, environmental policy analysis, water resources planning, and methods of benefit-cost analysis. His Ph.D. thesis is widely considered the first publication on contingent valuation, a method in wide use today to quantify environmental benefits and damages. Dr. Davis has served as an adviser to foreign governments, has served in faculty positions at several universities, and has served on the staff of Resources for the Future. Dr. Davis received his B.S. degree and his M.S. degree from the Ohio State University and his MPA and Ph.D. degrees from Harvard University.

LEO M. EISEL is a principal engineer at Brown and Caldwell in Denver, Colorado. Dr. Eisel has more than 29 years of experience with water rights and water resources. He is the former director of the Illinois Environmental Protection Agency, the Illinois Division of Water Resources, and the U.S. Water Resources Council. He is also a past president of McLaughlin Water Engineers in Denver. Dr. Eisel has served on several National Research Council committees and has served as a member of the NRC's Water Science and Technology Board. He received his Ph.D. degree in engineering from Harvard University .

STEPHEN W. FULLER is a professor in the Department of Agricultural Economics at Texas A&M University. Dr. Fuller's fields of research focus on transportation, marketing, and international trade issues, with an emphasis on the economics of Mississippi River waterway transportation. Dr. Fuller served on the NRC Committee on Freight Transportation Needs for the 21st Century. He is author of 280 refereed journal articles and reports that focus on agricultural transportation and marketing issues. Dr. Fuller has been honored five times by the Transportation Research Forum for his research by receiving the Outstanding Paper in Rural Transportation Award. Dr. Fuller received his B.S. and M.S. degrees in agricultural economics, and his Ph.D. degree in economics, all from Kansas State University.

GERALD E. GALLOWAY is vice president of the Enterprise Engineering Group at the Titan Corporation in Fairfax, Virginia. Dr. Galloway is a former secretary of the U.S. Section of the International Joint Commission. Dr. Galloway has served as a consultant on water resources engineering and management issues to the Executive Office of the President, the World Bank, the Organization of American States, the Tennessee Valley Authority, and the U.S. Army Corps of Engineers. Dr. Galloway is a former dean of the Academic Board (chief academic officer) of the U.S. Military Academy. Dr. Galloway holds master's degrees from Princeton, Penn State, and the U.S. Army Command and General Staff College. Dr. Galloway received his Ph.D. degree in geography from the University of North Carolina.

LESTER B. LAVE (IOM) is the Harry B. and James H. Higgins Professor of Economics and University Professor at Carnegie Mellon University. His fields of research include applied economics and public policy, safety goals for dams and other structures, and quantitative risk assessment. Dr. Lave chaired the NRC Committee to Review the Upper Mississippi River-Illinois Waterway Navigation System Feasibility Study. He is a member of the U.S. Environmental Protection Agency Science Advisory Board and the former president of the Society for Risk Analysis. Dr. Lave received his Ph.D. degree from Harvard University in economics.

KARIN E. LIMBURG is an associate professor at the College of Environmental Science and Forestry at Syracuse University. Her fields of research focus on the Hudson River estuary in eastern New York State. Dr. Limburg teaches a course in fisheries biology and is a co-convener of a seminar series in interdisciplinary courses in watershed ecology. She received her A.B. degree from Vassar College in ecology-conservation and biology, her M.S. degree from the University of Florida in systems ecology, and her Ph.D. degree from Cornell University in ecology and evolutionary biology.

ELIZABETH A. RIEKE is the Lohontan Basin area manager for the U.S. Bureau of Reclamation in Carson City, Nevada. Ms. Rieke is a former director of the Natural Resource Law Center, University of Colorado School of Law, and a former assistant secretary for water and science in the U.S. Department of the Interior. She has served as an associate (1987-1989) and as a partner (1989-1991) with the law firm Jennings, Strouss & Salmon. Ms. Rieke received her B.A. degree from Oberlin College and her J.D. degree from the University of Arizona.

RICHARD E. SPARKS is the director of the Illinois Water Resources Center at the University of Illinois, Urbana-Champaign. Dr. Sparks' fields of research include biological monitoring for pollution control; restoration of degraded aquatic ecosystems; and ecology of large floodplain rivers. He is a member of the American Fisheries Society, the Ecological Society of America, and Sigma Xi. Dr. Sparks was a member of the National Research Council's Committee on Aquatic Restoration and Committee to Assess U.S. Army Corps of Engineers Water Resources Project Planning Procedures. He received his B.A. degree from Amherst College, his M.S. degree from the University of Kansas, and his Ph.D. degree in biology from the Virginia Polytechnic Institute and State University.

SOROOSH SOROOSHIAN (NAE) is a distinguished professor and the director of the Center for Hydrometeorology and Remote Sensing in the Department of Civil and Environmental Engineering at the University of California, Irvine. His fields of research include surface hydrology (with an emphasis on precipitation runoff modeling), the hydrology of arid and semiarid regions, and related water resources management issues. He has served on several National Research Council committees, including a six-year term as the chair of the NRC Global Energy and Water Cycle Experiment (GEWEX) Panel. Dr. Sorooshian was elected to the National Academy of Engineering in 2003. Dr. Sorooshian received his B.S. degree from

California State Polytechnic University and his M.S. and Ph.D. degrees from the University of California, Los Angeles.

STAFF

JEFFREY W. JACOBS is a senior program officer at the National Research Council's Water Science and Technology Board. Dr. Jacobs' research interests include policy and organizational arrangements for water resources management and the use of scientific information in water resources decision making. He has studied these issues extensively both in the United States and in mainland Southeast Asia. Since joining the NRC in 1997, he has served as study director for 13 study committees. He received his B.S. degree from Texas A&M University; his M.A. degree from the University of California, Riverside; and his Ph.D. degree (all in geography) from the University of Colorado.

JOSEPH R. MORRIS is a senior program officer with the National Academies' Transportation Research Board (TRB). On the staff of TRB's Studies and Information Services Division since 1983, Mr. Morris has participated in studies of freight transportation, highway safety, transportation finance, highway design standards, and transportation and air quality. He received a master of city and regional planning degree from Harvard University, an M.S. from the University of Chicago, and a B.A. from Oberlin College.